TABLE OF CONTENTS

The first edition of this book did not have a Table of Contents so it does not appear where contents are normally located.

Introduction i
A Funny Thing Happened To Me On The Way Through Life 1
The Risley-Whipple Barn 3
Have You Ever Been To A Barn Dance? 5
Courage Through The Storms Of Life 7
Zipper, The Wonder Dog 9
The Christmas Tree 11
The Escalating War 13
Faith Sustains Me 15
The Human Touch 17
Dirty Children 19
The Big Horse 21
The White Owl And Baby Skunk 23
The Hook 25
The Silo 27
I'm Cool, Man 29
Saved By The Town 31
Wild Horses 33
The Risley-Whipple Outhouse 35
More Out-House Stories 37
Sailing Down Lake Road 39
Grandpa's Wisdom 41
Did You Ever Hear Of A Groundhog Thresher? 43
Fire And Barns Don't Mix 45
Cooper Shop Barn 47
The Hooded Rambo 49
Halloween On the Farm 51
The Glass Is Broken 53
The Sports Arena 55
What Keeps Me Going 57
Camp Wanake Inpost 59
Camp Wanake Outpost 61
Please Be Careful On The Farm 63
Grandpa And Grandma Whipple's Barn 65
The Horse Trainer 67
Lost Religion 69
Holy Barns 71
Model T Fords 73
What A Rare Gift 75
Did You Ever See A Boat Barn? 77
Guess 79
Heina Lato 81

The Renaissance Barn 83
A Streetcar Barn Named Desire? 85
Dead Or Alive 87
Horses Should Not Make Pigs Of Themselves 89
I Speak Of History 91
I Was Built With Marriage Marks 93
The Barn Consultant 95
I Am A Civil War Survivor 97
Tales Of Conneaut Creek 99
The Unbelievable Fours 101
My Dream Journey 102
About The Author .... Inside Back Cover

Library of Congress
TXu1-293-014
March, 2006
ISBN 978-0-9796143-0-9

INTRODUCTION

Barns are not human, nor do they desire to be so after what they have seen and heard from many humans. In spite of all that, they still love us and enjoy our being around.

Actually, barns are built for very specific purposes, such as to house horses, cows, sheep, goats, and sometimes, even chickens, ducks or turkeys. They are also a convenient home for cats and dogs that the owners do not want in the house. They store hay, grain and farm utensils. Although they were not built for this purpose, they are a wonderful place for children and some adults to play, especially, on rainy or snowy days.

But barns are more than wood, nails, roofing, and glass. I like to believe that they have the ability to see without having eyes, hear without having ears and experience feelings that are close to human emotions. One might consider that they might even have a heart of love.

Most significantly, I believe that the older, broken down, decaying barns are much like aging people filled with memories. With their memories, similar to aging people, they keep hanging on with faith, perseverance, courage and humor.

Any similarities to anything that might have happened to any members of my family, my friends, or myself are purely coincidental. Furthermore, I vehemently deny that I can be personably identified with the occasional icons found on the margin, except under oath. I use to tell my mother that I wanted to write a book about my childhood and she made me promise not to publish it until she was safe in Heaven away from the embarrassment. I have honored her request since she arrived in Heaven in 1991. I must admit that even as I read of that rascal, indicated by the icons, I am reminded of cartoon characters called Calvin and Dennis who cover two generations.

My deepest appreciation to Jan Page of Mt. Vernon, Ohio, a dear friend and a wonderful professional photographer who freely gave me many photos of old barns for this book.

I also wish to thank another dear friend, Lynette Host, who graciously has been reviewing my writings for sentence structure and omissions, of which there are many.

I would like to present this book in memory of my late wife, Dorla, who patiently waited while I took photos of many of these barns over the years anticipating that someday I would write this book. I would also like to present this book in memory of my father and mother, Paul Sr. and Hazel Whipple, who survived to live long lives, in spite of their children.

Paul Whipple

A Funny Thing Happened To Me On The Way Through Life

You humans may think that you are the only ones who ever have funny things happen around you. Well, let me tell you of a couple of funny things that have happened around me.

I remember my owner, who was about sixty-five years of age at the time, had walked through the barnyard hundreds of times and that young bull had always ignored him to the extent that my owner never worried about turning his back to that young bull. That was a big mistake.

I have no idea what got into the brain of that bull. Maybe he thought he would have a little fun, or maybe he was mad at my owner, or maybe he was just being a bull but one day when my owner's back was to him that bull lowered his head and came charging at full speed. He nailed my owner right in the buttocks flipping him end over end through the air.

My owner could have been severely injured except for the fact that he landed right on top of a fresh manure pile that was just outside my door. I have no idea whether or not that bull planned that in his charge but it turned out to be a good place to land. My owner had no broken bones but he was mighty lame for a couple of weeks. My owner had a habit of spewing a few expletives at his animals even though I never heard him express those same words at his fellow human beings. I can only imagine what he must have called that young bull.

You will think that I am full of bull, but it just so happens I have another funny bull story to tell you. My owner's grandson, Sherman, who was about sixteen at the time, was bringing in a young bull who had been tethered out near the road which was about one hundred fifty feet from me. The young bull was on a fifteen foot chain that Sherman had wrapped around his hand. I suppose that the bull suddenly had the thoughts of supper in his mind because he decided to make a run for me at top speed. A small door had been cut into my big sliding doors which certainly was not big enough for Sherman and the bull, especially, at the speed they were going.

The chain tightened on Sherman's hand and he could not get loose. By this time Sherman was screaming bloody murder alerting the whole family. Sherman's feet were hitting the ground about once every fifteen feet. All Sherman could think of was that little door that wasn't big enough for the two of them. Just as the bull reached my door he suddenly stopped, much to the relief of Sherman and all his family.

It is humorous events such as these events that keep me going and still chuckling at the memories. That is the only way to survive. So laugh a little and be sure to see the funny side of life.

The Risley-Whipple Barn

Would you believe that my owner, P.J. Risley, once moved me about 300 yards from the back of his house to where I finished my days near the Whipple home? He, with the help of some neighbors and relatives, used a series of block and tackle, wooden rollers and a team of horse to accomplish the task. It was a very slow process but I made it to my new location all in one piece. The horse barn and the hay loft above it, that was at my south end, were built later and was the first portion of my frame to disintegrate and be torn down. I loved it in my new location where I was so close to the Whipple family who moved into the little house next to me about 1931. It was the first of four homes that P.J. built upon the property. Paul Sr. and Hazel were married in that little house in 1913 so it was coming home for Hazel.

The Whipple family comprised of Paul, Hazel, and their four kids, Graedon, Sherman, Phyllis, and Paul Jr. (Oh, how he hated to be called that, especially when the older children would teasingly call him "Paul Junie"). There were always more people around with Hazel teaching piano lessons. In addition, Hazel was always bring home stray children and Paul Sr. was always bringing home stray dogs. There would be times that I would be harboring as many as five dogs that Paul had rescued from the dog pound. His question to everyone he met was "do you want a nice dog?".

The family stored their ugly orange picnic table and wooden horses on my main floor. Whenever someone would come and get that table I knew that there was either a family picnic in the backyard or an ice cream social for the Amboy Methodist Church in the front yard. I knew it meant the excitement and laughter of kids of all ages playing in my many rooms. There is so much that I could tell you what I saw and heard but I considered it a sacred privilege to be apart of the family and all their friends so I keep their secrets well. By the way, I understand that the old orange table and wooden horses has been passed down to the next generation and now resides at the home of Don and Mary Ann Little in Ashtabula where the family reunions continue.

After Hazel, the matriarch of the family, died at the age of ninety-six and the family had sold the homestead I knew that my days would be numbered because I was also getting frail and very old. So, the new owners tore me down but guess what? They built a beautiful new garage on my old foundation so there is apart of me that still lives, plus the Whipple family and their many loved ones and friends who still remember me with love as I remember them so fondly and smile.

Have You Ever Been To A Barn Dance?

Would you believe that I was once one of the most popular buildings in the County? It was because I was where the people, young and old, would come to barn dance. They would open my large front doors to let in the fresh night air and often hung Japanese lanterns outside to create the feeling of charm and festivity.

My wooden floor had become smooth from the years of straw and hay that had been trampled underfoot by farmers, horses and wagon loads of hay. The square dancers also helped to smooth out my floor so I was a great place to go dancing.

On the night of the dance the adults would arrange folding chairs along both sides of my main floor while the fiddlers would set up near the door to enjoy the coolness of the night. The steady drone of the caller was almost hypnotic as the girls in their bright dresses and white petticoats would whirl around the dance floor with their partners while the old folks would tap their feet, recalling the days that they were out there dancing.

The young children would sit up on the edge of my hayloft watching the young dancers with envy, dreaming of the day that it would be their turn on the dance floor. Young boys would also sit on the edge of my hayloft with their blushing sweethearts, occasionally stealing a "not to be forgotten kiss". If the younger children happened to see them, they would giggle and whisper excitedly.

Originally, the only light available was from globed kerosene lanterns that hung from various posts. The light that they shed was just enough to see when your eyes grew accustom to the dim light but leaving many dark areas soft enough to be romantic. I admit that I liked it better then when they installed electric lights. I know that the electric lights were much safer but the lantern lights were so much softer and mystical.

I can still hear the rhythmic beat of the dancing feet, the staccato clapping of the hands of the older folks, and the nasal drone of the caller as he called the Virginia Reel or Jimmy Crack Corn. Can you hear it too?

There is nothing to compare to the joy of square dancing in a barn and I was one of those barns that made it all possible and an unforgettable memory.

Courage Through The Storms Of Life

It takes courage to stand up against the blizzards and freezing rain of winter. Sometimes the hail upon my roof was so noisy that conversation was difficult. Sometimes the wind would blow through every crack of my sides causing little piles of snow inside. There would also be times that drifting snow would cover my doorways causing the silence to be almost deafening. But it was in those times that I was most appreciated as a shelter and refuge from the storms by my animals and rodents.

In the spring I would be drenched with numerous rains that did not give my boards time to dry out before another rain was upon me causing some of my boards to warp and shrink. But with the spring there also was the fragrance and beauty of the spring flowers and the hope of summer.

Then would come the summer with the heat of the sun beating down upon me, bleaching my outside walls a silver grey like the hair of older humans. It was always a joyous time when the fresh alfalfa and hay would be stored in my lofts and the grain bins would be filled.

Naturally, my greatest fear in the summer would be during the frightening electrical storms that so easily could destroy me and all the animals I sheltered. I remember one owner whose son would sit in the kitchen in their home nearby with a hatchet to cut the tether that held the horses. He would also hold a towel to cover the horses eyes in order to lead them out should I be hit with lightening. Sometimes the thunder would be so loud that my whole foundation would shake. When the storm was over there was often a rainbow in the east against the black clouds of the storm that had past. The air would feel so clean and fresh when it had been so sultry and heavy before the storm. As the last of the water would drip off my roof into a shallow puddle it had made and the sun would return I would thank God that I had been spared through another storm and all my animals were safe.

Life is not always easy but it is certainly not boring. God never promised that life would always be easy without storms that tax our faith and courage. I understand that the Bible says that God lets the sun to shine on both the good and evil and lets the rain fall on the just and the unjust. As much as you might like it otherwise are you not grateful that God does not discriminate against you when you happen to be evil or unjust?

Finally, God promises to be with us until the end of life. What more would you want? Be of good courage!

Zipper, The Wonder Dog

I know that you will have a difficult time believing what I have to tell you about Zipper, a mixed breed small hound dog owned by the Whipple family.

Zipper was free to roam inside of my walls, but he was not permitted to live in my owner's home. Zipper was always interested in what was happening outside, especially, if he heard human voices.

On various occasions, the family members were amazed to see Zipper peering out of my highest window that was, at least, twenty five feet above ground level and for most an inaccessible area. The only way that humans could reach it was by a vertical ladder that was made with wooden slats nailed to one of my main columns. The question for the family was "how did that dog get up there?"

The answer came one day when Paul Jr. happened to come in through my door quietly and saw Zipper actually climbing a twelve foot rung ladder that was at my south end which was above the horse stables. How he figured out how to climb a rung ladder no one knows.

Then he ran to my east side where there was a small opening to the east hay mow and jumped into the hay. The hay was sloping downward so Zipper climbed to the highest level and leapt to the flooring that was directly over where the hay wagons would come in through my doors. There he could look out my window. When he would become disinterested Zipper would then leap back onto the hay mow, walk to its lowest level and then leap five or six feet down to the floor. That was some amazing dog!

It was just surprising things like that situation that made my days and nights a little brighter and filled my memories with warm thoughts and a little chuckle.

The Christmas Tree

How do you like my permanent Christmas tree? Actually, it is not a Christmas tree but the remains of a windmill covered with ivy that is about as old as I am.

How I loved to hear that windmill sing night and day as it carried the storm winds off Lake Erie with the faint scent of marine life. Not only did I hear the whirl of the windmill blades but I also listened to the rhythmic thumping of the long piston that drove the generator. The generator provided electricity for the farm, including pumping water from the well. In its day it was considered the latest modern thing that revolutionized the farm.

When strong winds would blow, that windmill made such a racket that I thought its blades would fly off and the whole structure would collapse around me. But it, like myself, survived many years and many storms even though we are only remnants of what we once were.

There were times that it only hummed along at a lazy pace. Then there were times when there was no wind at all and the blades were still. In those times, we would all revel at the peaceful silence, listening to the lovely sounds of nature such as the chirping crickets in the hay or the peaceful, joyful March peepers that announced that spring had arrived.

Such beautiful memories! I pray that you also have beautiful memories. Please try to remember pleasant thoughts.

Photo By Jan

The Escalating War Games

I remember when Richard, who was about thirteen and his brother, Duane, was about eight, used my many rooms and hay lofts, as an ongoing war game. They used bales of hay to create tunnels and secret passages. Their weapons of war were shelled corn cobs that they had soaked outside in the horse trough. Soaking the corn cobs added to their weight, accuracy in throwing, evidence of a direct hit, and best of all, got the enemy wet.

The war game was doing great until one day Richard decided to escalate the war game with a brand new weapon of mass destruction. It was the terrifying, enemy destroying, greatly feared weapon of war- the pea shooter. A pea shooter looks much like a drinking straw with the diameter big enough to shoot dried peas through it by holding the pea shooter to one's mouth and blowing a pea through it.

With the introduction of this new weapon of war, Duane was losing the battle until he made a very lucky shot with his wet corn cob. He happened to make a direct hit on Richard's pea shooter as Richard was about to blow a stinging dried pea out of his mouth.

The result was that this newest weapon was not only broken but Richard's front tooth was also broken. The only consolation for Richard was that he was the only boy in his school who could spit through his teeth.

And that was the end of the battle of the wet corn cobs.

Faith Sustains Me

You may wonder what keeps me going or even why I would want to keep going? I'll gladly tell you the answer. It is my faith and trust in God our Creator and Father who is not only in charge of the universe but maintains my existence as well as yours.

You look at me and think of me as dead, decaying and worthless. God does not consider me worthless even in my condition. I am no more dead than dying people in your community who are very much alive, even though they may be dying. Please remember that dying is very much a part of living. Death occurs when the soul departs the body. My soul has not yet departed the barn I once was.

Though I may not possess the strength and usefulness I once had, I am very much alive and useful to some of God's creatures. For instance, I can still provide a home for rodents, birds, bats, and even small animals who may raise their young within my sheltered walls. The song of the cricket still sings in my moldy, decaying hay that still clings to my drooping walls. Barn owls still watch for the swift dart of a mouse across my floor mindful of their next meal.

Aging is not an easy journey but with faith in God to watch over our very existence we have only to endure with patience and understanding. I pray that you trust God to watch over you as well.

The Human Touch

Wretched as I may appear and lonely as I may be, I am still at peace. I served my time well. I provided refuge for many cows, new born calves, stomping horses and an endless number of cats and kittens too numerous to count. Occasionally, I would house a dog who loved to bark his resistance to being separated from his human companions.

My happiest moments were when my owner or the members of his family would regularly visit me to milk the cows, feed the animals and clean the stables. Even more exciting were the times that my owner's children would come into the barn. You could never be sure what they might do.

They might jump in my hay loft, or play hide and seek in my many rooms. They would often squeeze a cow's udder squirting milk in the cat's face. The cat would stand on its hind legs to better catch the stream of warm milk with its mouth wide open and eyes closed. Afterward, the cat would sit quietly licking off the milk that had splattered its fur.

I recall two young boys who played a game, the name of which I cannot imagine. They would start a foot race out though the buggy area, then streak through my big doors, out through the back door, and then back in through the cow stable. The first one to get back to the cow would grab one of the cow's udder and squirt the loser. I suppose that you might say that the real loser in that race was the old cow but she never seemed to mind. Who knows, maybe she liked the excitement. At least, she never kicked the boys in protest.

CHEW
THE BEST

Dirty Children

Before the days of TV, children invented their own games, some of which were very messy. Occasionally, someone would leave me an old newspaper that had a humorous cartoon called "Why Mothers Become Grey". The things that have happened around me would cause any mothers' hair to become grey.

For instance, there were two young boys who came to their mother and were absolutely covered with mud from head to foot. If you have ever been at the back of any barn that has cows in it, you know that just outside the stable doors, where the cows wait to come in for milking, there is a terrible muck which is usually four to six inches deep. Well, these two boys had been behind me, entertaining themselves by hanging on various cows tails and being dragged through all that mud. A mother of one of the boys took them outside and washed them down with a garden hose.

The son of one of my owners, and his friend who lived nearby, were two of the most ornery boys I ever remember. Near me there was some molding sand pits created by men who shoveled out this very fine sand and loaded it into dump trucks. When it rained heavily, the sand pits would fill up with water, waist high on the boys. Even though Lake Erie was less than two hundred yards away, these boys would wallow in that sandy water that was sixty per cent water and forty per cent molding sand in suspension. When they would go home they had yellow skin, yellow hair and, of course, yellow clothes.

The property next to me was owned by the Cummings Canning Factory who raised huge acreage of tomatoes or pumpkins. Whenever the tomatoes were ripe these same two boys would have to have, at least, one good tomato fight. They would come home looking like they had been in a terrible accident and were bleeding to death.

Once, when the pumpkins were ripe, they ventured over into Cummings field where there was a small stream. The boys decided to try to dam up the stream with pumpkins. They lugged two wagon loads of pumpkins down to the stream but finally concluded that you can't dam up a stream with round pumpkins.

A more disgusting way they entertained each other was to have a cow pie fight. They would each get a green stick about three feet long, stand about thirty feet apart, bend the stick into a wet cow pie and flick it at each other. If you don't know what a cow pie is, ask anyone who grew up on a farm. I never did figure out what their two mothers did when they came home after such an assault.

They also loved to pull a trick on any city kid who happened to come around. They would shove a stick into a fresh cow pie and then poke it toward the city boy's eyes. Invariably, the city boy would grab the end of the stick ending up with a handful of that filthy stuff. It was things like that which caused country boys to learn to run fast.

And it was such kid behavior that kept me from being bored and with great memories. By the way, one of those two boys grew up to run a good factory and became a fine leader in his community and in his church. The other became a preacher. His claim was that he had to become a preacher as an act of penitence.

The Big Horse

My owner owned the biggest horse I ever saw. He bought it for $50 at an auction because no else wanted it. It was so big that he had to build a special harness to fit the horse.

But the most amazing thing was that my standard horse stall was too low for the horse to raise his head. So, Alden, my owner, tore up the floor and dug out the stall 2 feet deep so the horse could lift his head normally. Believe it or not, Alden was able to hitch up that big horse to another horse as a team which was very funny to see and was the amusement of the nearby farmers. But Alden seemed to get them to work together in spite of their differences in height.

Outside my walls there was a fence for the cows to be corralled into my barn. Alden would have to go out into the woods and outer pasture to round up his cows for milking. Why his cows would not come in like ordinary cows I will never know.

So Alden would ride that big white mare bareback to round up the cows. To get on the horse Alden built steps along-side the fence to climb on the horse's back.

One day he had gone to town and gotten drunk. Somehow, he returned home safely and was able to put the bridle on the horse. He had the horse trained to come to the fence where he could climb on its back. But this time he stepped out from his perch at the top of the fence and missed the horse, landing in a heap on the ground.

Stubbornly, he kept repeating this act until his wife called their son, Russell, in town frantically begging him to come out before his father killed himself falling off that horse. Alden was stiff for at least a week after that episode. It didn't stop Alden from getting drunk on occasion but he never again climbed on the back of that horse unless he was sober.

The White Owl And Baby Skunks

Many years ago a large snow white owl roosted in my rafters. It was the biggest owl I had ever seen. I had no idea why a snow owl would be in northern Ohio when their normal habitat would be in northern Canada or Alaska.

My owner decided that he would like to capture that beautiful creature alive. I can't imagine what he thought he would do with it after he caught it. He was a Boy Scout leader at the time and called some of his scouts to help him catch the owl.

Well, my owner sort of caught the owl. The truth of the matter was that the owl caught my owner. That owl clamped a talon right through my owner's right hand. They actually had to take a pair of pliers to remove the talon from my owner's hand. The owl flew away when it was released. I hope it was able to tell its grandchildren of the time that it caught a human. I don't recall my owner ever trying to catch another owl.

Let me tell you of a more comical incident that took place when that particular owner still lived there. A mother skunk had a number of baby skunks in a hole under my foundation. Something may have happened to the mother skunk because she was nowhere to be found.

My owner was standing around with several men discussing what was the best way of getting those baby skunks out from under my foundation and into a barrel. There was the question as to whether or not baby skunks had the ability of squirting their odoriferous perfume and they were hesitant to find out.

While they were talking, the owner's five year old son simply walked over and picked up each one of the baby skunks and dropped them into the barrel, much to the astonishment of the men who were standing there.

I want to tell you that there was a smiling, triumphant little boy and a bunch of sheepish looking men that day.

The Hook

Children were always the most joy in my life. Some of the things that they would do you wouldn't believe. Other things they would do you don't want to know. But let me tell you of one hilarious incident that took place on my main floor.

This young boy, who was about nine years old, happened to be in my main area with four of his girl cousins, one of whose smile would make his knees weak. It was one that he would have liked to be a "kissing cousin" but she only teased him with her smile.

They dared him to ride the block and tackle that hung from my rafters to hoist bales of hay into my loft. Wanting to impress his "wanna be kissing cousin" he accepted the dare.

So he climbed onto the hook and the four girls pulled him closer and closer to my ceiling.

Suddenly, there was the distinctive sound of cloth ripping as the hook wiped out the entire crotch of his pants, obviously, exposing his undershorts. The girls were laughing so hard that they just kept him up there in the air, much to his own humiliation.

When they finally let him down he, embarrassingly, backed out of the area and ran to his cousin's mother who pinned up his pants enough so that he could ride his bicycle home.

You can't imagine barns ever having fun? Well, that was a fun day for me that I will never forget!

I was told that this young lad grew up to be a preacher. One evening he was in another barn playing games with his youth group. He ran to the center of the group to retrieve a ball. When he bent over to pick up the ball he ripped out the back of his pants again, much to the hilarity of the group. Needless to say, he quit playing the game. What's with this guy?

Photo By Jan

The Silo

Isn't that a beautiful picture of me along with my silo? It was taken by an artistic professional photographer. Were you aware that few barns still have their silos intact?

Filling my silo was always an exciting time . My owner would rig up his tractor in a stationary position and hook a large pulley belt to the auxiliary power on the side of the tractor's engine. On the other end of the belt he would attach an ensilage chopper and blower.

The flying blades of that chopper would cut up the green corn stalks as fast as the farm hand could feed in the corn stalks. The blower was so powerful that it would blow that chopped up corn stalks forty or fifty feet through a vertical tube to the top of my silo where the end of the tube bent down in order to drop the ensilage into my silo.

The corn stalks had to be green so that the ensilage would ferment. Did you ever smell fermented ensilage? It is such a sweet and delicious odor. I understand that some farmers even mixed molasses to their ensilage to increase its sugar content.

There were iron bar steps mounted inside and outside the silo so that a farm hand could climb down inside the silo to spread the ensilage evenly using a shovel or hay fork. Some of the silos had a movable pipe so that the farmhand could spread it around more easily. I remember years ago when a six year old neighbor boy would be hired to climb the outside ladder of my silo and remove the door down to the level where the ensilage had settled. He would then shovel the ensilage out the door and down the chute to the ten cows in their stalls below. For his daily effort that generally required two hours of work he would receive the handsome reward of two pennies. Outside my barn, some farmers were saying that in Hawaii they grind up the rinds and leaves of pineapples to make ensilage. Do you suppose that it would make the cows produce milk with a pineapple flavor?

The odor of well fermented ensilage would filtrate my whole barn. Sometimes it would permeate the farm house nearby, especially, noticeable at meal time when it interfered with the taste and flavor of the food. It smelled intoxicating. Do you suppose that if a cow ate too much it would get drunk? I've been told that young boys have tipped over standing sleeping cows. If that is true do you suppose that you could knock over a slightly inebriated cow?

I'm Cool, Man

In the days before air-conditioning the coolest spot during the hot summer months was inside my double walls lying in the damp sawdust. I may have been small, and maybe not even classified as a barn, but I was a very important outhouse on any farm. I was the ice house.

During the winter months my owner would go out on the frozen lake behind the barn and saw out huge chunks of ice with a crescent shaped piece of iron that had jagged teeth on it and a wooden handle. Then, with a pair of ice tongs he and another man would drag out that chunk of ice and haul it onto the bank of the lake. Then they would load it onto a bobsled that was hitched to a team of horses.

When the bobsled was heavy laden with chunks of ice, they would bring it to my upper door and carefully transfer the ice slabs into my storage area. Then, they would be sure that there would be sawdust packed all around the slab, and especially, between every layer so that they could retrieve the ice easily when they needed it in the summer.

So on hot days during the summer I was a haven for children who would crawl up the ladder outside my door and then climb into that unbelievable cool refreshing damp sawdust. Lying there with the cool vapor of the top layer of ice surrounding them they thought they were in paradise.

Every few days, during the summer months, my owner would come out of the farmhouse with a big coal bucket and ice pick. He would chip off large chunks of ice, place them in a burlap bag, and then break them into smaller usable pieces with a sledge hammer. The final step was to wash off the sawdust at the pump that was in the backyard.

The family would use the ice to keep perishable food cold in the ice box, or to make home made ice cream and even to cool a glass of lemonade. I don't recall anyone getting sick from consuming that ice. You see things were a lot cleaner in those days and life was a lot simpler. We use to refer to them as the "good old days".

By the way, please remember that being small doesn't make you less important.

10 7'02

Saved By the Town

Do you see how beautiful I look today? Just a few years ago I was somewhat of a dilapidated barn with very little hope of survival. The plans to demolish me and make way for a housing development were already in progress.

But some wonderful things happened when the president of the Hudson Heritage Association, Attorney Douglas Whipple, headed up a massive effort of many organizations and local government officials of Hudson to raise the funds necessary to invest in a private organization with a dream. They invested in the Case-Barlow Farm Inc. to purchase the farm and to turn it into one of the finest community projects to be used by all the citizens of Hudson and, especially, their children for generations to come.

Once the farm was purchased the Land Legacy Committee of Hudson, chaired by Dick Bashore, acquired the sixty-six acres of adjacent pasture land and developed it into four baseball fields, six soccer fields, a family picnic pavilion and walking trails. Talk about community spirit! But let me tell you more about myself. I was built in 1891, well over a hundred years ago. Through the years as I was starting to show some wear and tear, the frugal Barlow family would patch up the holes on my outer wall with Ohio licenses. It is still possible to read the year of those license plates even over my beautiful new coat of paint.

My historic farmstead is where the Civil War abolitionist, John Brown, played as a young boy with his best friend, Lora Case. Meanwhile, the adult members of the Case family illegally harbored runaway slaves in the fields on the slaves' long, hard journey to freedom. My picture was taken at the annual Farm Fest, which draws hundreds of people. The day before they repainted me, Sherwin-Williams Co. provided paint brushes, caps, aprons, and paint for all the children in Hudson to paint a barn. The children were permitted to paint anything they wanted such as letters, faces and whatever came to mind. It was a glorious fun filled day with a fascinating display of children's art, a day that the children will never forget. The non-profit organization has also fixed up the farm house and replaced my silo roof with a replica of the original.

I boast two eras in agricultural history. At the west end my original milk house is still attached. Farmers learned that this was very unsanitary, so the old milk house was replaced by a "modern " milk house in the 1940's which is not physically connected to me. The building blocks of the new milk house are made of ceramic tile. This may be the last milk house of its kind still standing in Ohio.

I, along with the rest of the farm, have been saved because of the vision of the leadership in Hudson and the generosity of the Hudson people. Thank you, Hudson!

Wild Horses

For a number of years I was the proud keeper of two beautiful, giant Clydesdale horses whose names were Doc and Dowdy. I'm sure that Doc wouldn't mind being called by that name but what respectable horse would ever like the name Dowdy? They were used to haul logs out of the forest behind me that involved some really difficult hills. They were so strong and adept at working in the woods that they could actually climb over fallen trees while dragging a log behind them.

My owner's two young boys would ride those great horses bareback through the woods. But they had one slight problem. Whenever Doc and Dowdy decided that they had enough for the day, they would simply turn around and race back to the barn totally ignoring the boys attempt to control them. The boys simply had to let them go.

My door opening was a normal height but too low for these two Clydesdales. So when they went under my door they would lower their heads while their backs were only about one foot lower than the door casing. Doc and Dowdy didn't quit running until they were back in my main floor. So these two boys, to avoid being decapitated when they passed through my door opening, would simply reach up, grab hold of my door casing, slid off the back of the horses and drop to the ground.

I remember when another one of my owner's had been threshing at a neighbor's farm. So when he was ready to come home his horses were as anxious to get home as was their owner. So they were running a fast clip down this slight incline. The wagon was empty which made their running quite effortless.

My owner never knew whether he may have pulled the reins to the right or what else might have happened, but suddenly, these two horse veered off the road and ran on both sides of a tree. My owner was thrown into a ditch but was not hurt. The horses and the wagon escaped harm but the tongue of the wagon was demolished. Would you ever wonder if those two horses might have winked at each other, snorted a little chuckle and just might have had a grin on their faces?

I also recall that out in my pasture a young lad, who had never ridden a horse, was hoisted into the saddle of this horse. The owner slapped the horse on its rear end and yelled to the horse, "Giddyap"!

The horse took off like a race horse, flying across that pasture right toward the fence. The horrified boy clung to the horse for dear life, not knowing whether the horse would try to leap the fence, or throw the boy off when the horse reached the fence.

Luckily, the horse stopped just before they reached the fence and the frightened, trembling boy slid off the horse to the ground. I thought that was a very cruel and dangerous trick to pull on that young lad who had never ridden a horse. He could have been seriously hurt, or even killed, and especially, so terrified that he might never have ridden a horse again. Fortunately, the boy never let that first experience keep him from riding a horse whenever he got the chance.

Horses are wonderful animals but be aware of their strength and unpredictability.

The Risley-Whipple Out-House

I know I am not a barn but you have got to admit that I was one of the most essential buildings on any farm. Please excuse the image of Paul Jr., at the age of sixteen, standing in front of me blocking your view of me. But this photo is the only known picture left of me. I was built by P. J. Risley who built the barn nearby, the garage next to me, the farmhouse in which Hazel and Paul Whipple Sr. were married, the house and barn next door and the two cottages on the bank of Lake Erie. He was a busy man!

I happen to have two adult seats or holes but I never recall two adults coming in and sitting together. But there were times when a mother and her child would come in and talk.

Did you notice the lattice work made out of tree limbs that created somewhat of a modesty screen, especially in the summer when the wild cucumber vines would block the view from the outside? Often my visitors would sit with the door open to avoid the stifling heat and an obnoxious odor that wasn't my fault.

I hate to tell you this but you need to know the reality so that you will appreciate your indoor plumbing. There were rat holes around my foundation that led down to that stinking mess. On windy days those rat holes would create such an updraft that you had to carefully wad up the soiled toilet paper and throw it down the hole or it would blow back up toward you.

Let me tell you of some of the hilarious things that took place when the Risleys and the Whipples lived at the farmhouse. P.J. had removed the pot belly stove from the living room in the late springtime, but he had left it halfway between the house and the garage with the metal stack sitting on top of it. His wife, Agnes, came out to visit me in the early evening and forgot it was there. As she was returning to the house she saw this dark figure suddenly before her. Thinking it was a man she socked it with all her might and heard the smoke stack go crashing to the ground. Lucky she did not try to punch that fella in the belly!

Sherman was about sixteen when he decided he would produce some creative artwork on the piece of linoleum that was nailed to my back wall to keep the wind out. He drove six big spikes through the linoleum from my outside wall. On each spike he hung a corn cob with the size determined by the family members' status. Paul Sr. naturally received the largest cob and Paul Jr., the youngest in the family, received the smallest. He then added each person's name on a piece of paper above the spike and secured the names permanently with a coat of shellac.

A couple of weeks later, Sherman had his cute girlfriend come for a visit. The instant she stepped out of the car, Paul Jr. grabbed her hand and dragged her out to my place to show her Sherman's great art work, much to Sherman's embarrassment.

One time, Phyllis decided to visit me wearing only her underpants and a gym blouse that had a flap of material that went between her legs and buttoned on the front of her blouse. So, Sherman and Paul Jr. each got their cameras and hid along both sides of me . When she came out and saw those two cameras she ran for the house but not fast enough. The boys got the picture with the blue tail flying high behind her.

5 26 '07

More Out-house Stories

Living on busy Route 14 just north of Deerfield next to a now vacated country store, I heard a lot of out-house stories which I would like to share with you. They are mostly true.

For instance, I heard of this young man who had gone to visit the cottage of his fiancee's parents at DeRuyter Lake in upper New York state. Having felt the call of nature at dusk, this young man went up the hillside behind their cottage to their out-house. Upon returning to the cottage he recommended that the father use a wood file and sand paper to smooth down the hole that had caused him great pain.

The whole family howled in delight, and ceremoniously, marched him back up to the out-house to show the young man the latest modern invention. The father had installed a modern toilet seat over the hole but it had been up against the back wall so the young man had not seen it.

Another family had five young daughters. One day, the mother removed the black fur collar from a winter coat and was ready to dispose of it. But the father had an idea how to have some entertainment with it. At dusk he went out and tacked that fur collar on the rim of the out-house seat. When the first young daughter went out to the out-house he called all the other members of the family to watch out the window. Moments later, the girl came shrieking out of the out-house with her panties around her ankles while the rest of the family roared with laughter.

This next story is not so funny. I know of a lady who was one of the most lady-like humans I ever met. She told of a time when she was out west in a desert area. She and her husband had stopped at a lonely gas station where there was only an out-house out back. When she had finished her business she stood up. By this time her eyes had adjusted to the dim light and she could see clearly. Curled up in the corner behind where she had sat was a rattlesnake! I suspect that she also made a fast exit.

I can't verify all the facts in these next two stories but the man who shared them told both stories with a straight face (which doesn't prove a thing). Anyway, one story goes that this old man was tired of having his out-house knocked over at Halloween by the boys in the community. He decided to hide in his out-house with a shot-gun to scare the boys away. Somehow, the boys found out that he was in there so they sneaked up behind the out-house. They pushed it over with him inside so that he was trapped inside with the door on the ground. He was so enraged that he shot his gun out the hole but by this time the boys were already out of range, laughing gleefully.

Another Halloween story, which I can't verify but it sounds feasible is about the man who was tired of having his out-house tipped over. So, after dark he went out and moved it three feet forward so that when the boys came running to knock the out-house over they fell in the hole. Yuck!

By the way, did you know that we were also known as a "privy" meaning a private, secluded place? No matter, we had an important part to play on every farm or rural home.

Sailing Down Lake Road

I am known as a carriage house, but most people prefer to call me a buggy barn. Most buggy barns are attached to the regular barn, but in my case, I was a separate building.

Keith had a buckboard buggy in my place. It was a lightweight, four- wheeled buggy that held two passengers on its seat with a small storage space in the back to carry groceries, etc. Most women preferred it to riding a horse into town in the olden days. It was pulled by a single horse who fitted between two shafts to which the harness was attached.

This particular buggy had remained for years in my place since the owners no longer had a horse. The automobile had taken its place as the most popular means of travel.

So, Keith had the creative idea of building a sail-buggy. First, he removed the shafts from the front of the buggy. Then he drilled a large hole through the floor above the right and left front axle. He then attached a piece of rope to each axle with the other end running through the hole up to the buggy seat for steering.

Then Keith cut a hole in the floor behind the seat large enough to insert a six foot mast. On the mast he rigged a stiff canvass sail five feet high and four feet wide.

Needing a co-pilot to assist steering and, I suspect, to take half the blame if he got into trouble, he invited his friend and neighbor, Paul, to join in its maiden voyage down Lake Road. Not being able to tack the buggy against the prevailing westerly wind, they pushed the buggy up a small hill to get a better start. Then they hoisted the sail and away they went, lickety split, down that road with the metal wheels chattering on the concrete and the buggy swerving from one side to the other, since the steering mechanism wasn't too great. Drivers in cars, seeing them come flying toward them, would pull off the road even though they were a long distance away.

The buggy had no brakes so the only way they could stop the buggy to let the cars pass was for Keith, who sat on the right, to pull sharply on his rope sending the buggy into an adjacent field. I never heard what some of the drivers in those cars said, but I could imagine.

In their Junior year in high school these two boys had a very unusual accident. On a very dark night, Paul was riding down that same hill as fast as he could on his bicycle trying to get as far ahead of his mother in her car, which she was driving to another neighbor's home. As he looked back to see his mother's car, he heard a frantic yell. It was Keith on his bicycle riding in the opposite direction to his sister's house. They crashed head on, destroying both bikes and causing injuries to both boys, especially Keith, who had a brain concussion. The center post went through his lower lip, breaking some of his teeth. In those days, one did not consider going to the hospital except to die, so the doctor came and patched up both boys at home. Keith's pupils would not dilate for a whole month so he had to remain in total darkness that entire time.

The same night of the accident a school mate was killed up on Rt. 20 when he was hit by a car. The rumor spread throughout the community that it was Keith and Paul who were killed.

In spite of their many harrowing adventures both Keith and Paul are still going strong in their early eighties with many stories to tell to their children and grandchildren. The Lord, in all his mercy, has watched over them again and again.

Grandpa's Wisdom

Sorry, there is no photo of a barn on this page because you already have three photos of me. I am the Risley-Whipple barn. The old photo is significant because it shows the three persons involved in my latest story. Sherman is on the left of the wagon while P. J. Risley is on top of the hay wagon and Paul Jr. is on the right. I want to tell you about an interesting conversation when my owner was mowing away hay in my loft with these two young men.

Since my owner had stacked the hay on the wagon he best knew how to unload it. Besides, he never let the boys drive his team of horses. So when they were in the barn, P. J. was on the wagon and tossing up huge fork fulls of hay to Paul Jr. whose job was to pitch it to Sherman who was mowing the hay away in my loft.

Hay has to be mowed away in a circular fashion so that when it is removed it can be easily accomplished by reversing the pattern in which it was mowed. Otherwise, it becomes very tangled and difficult to retrieve.

Paul Jr. was having a difficult time struggling under the weight of the fork full of hay that he received from his grandfather. He was staggering around in circles in his effort to get the work load of hay to his brother who was mowing it away.

Finally, P.J. who liked to chew tobacco called to Paul Jr. He then spit out some tobacco juice to be sure that he had his grandson's attention. He then said, "Son, I want to tell you something that I hope you never forget. Whenever you are working always remember to KEEP YOUR BUTT BEHIND YOU." (I must admit that was not the word he used).

So he taught Paul Jr. to plant his rear end against the center post, receive the fork full of hay on his right without losing the momentum of his grandfather's pitch and swing to the left releasing the hay to Sherman.. It worked like a charm.

That young man is now a man in his eighties. Throughout his life he has always tried to remember the words of his grandfather in whatever task he faced. In whatever he has ever done he has tried to remember, "ALWAYS REMEMBER TO WORK WITH YOUR BUTT BEHIND YOU."

5 26 '07

Did You Ever Hear Of A Groundhog Thresher?

Before the time of the giant threshing machines, which sat outside the barn and required a number of men to operate, there were small threshers about 2- foot square and 3- foot high that usually sat in the middle of the main floor of a barn. It was called a ground hog thresher. I never did understand where it got such a name. There were two kinds, one that was hand driven which had to be mighty tough work and the horse driven threshers. The single horse walked on a tread mill to provide the power. Believe it or not, the one on my main floor was driven by a team of horses who went around and around in a circle. Do you ever wonder if they might have questioned where they were going?

I happen to be a double barn for the very purpose of threshing inside. On rainy or wintery days when it was too miserable to work outside my owner would hitch up old Bill and Nell and drive them around and around the circle attached to that thresher. The thresher would spew out the grain in one direction and the chafe in a pile in the other direction. This pile of straw then had to be forked over to the loft in the other half of the barn. It was a slow process but, in those earlier days of farming, everything took a great deal of time. That is what made farmers so patient.

In order to avoid a supporting wall or vertical beams in the area where the two barns were connected the farmers developed a swing beam that was larger in the center and smaller on the ends. Somehow this design would carry the weight of the roof of the two barns, but don't ask me how. All I know is that I have one and it still works.

5 26 '07

Fire And Barns Don't Mix

I am an unfortunate barn that happened to catch on fire. However, I was blessed in that I was not totally consumed by fire, as are most barns that burn due to the high content of combustible materials such as hay, straw, or stored inflammables such as gasoline or oil. Please be careful around any barn because we are so vulnerable to fire.

Most barn fires are caused by lightning, spontaneous combustion, electrical shortages, carelessness, and worst of all, arson. I happened to be a victim of lightning.

Isn't it amazing that my entire frame of oak timbers survived? My sill and uprights are sixteen by sixteen inch oak timbers while my support beams are eight by eight inch oak beams. My owner could build a new wall around my frame anytime in the future. I hope that he does because I feel so naked and useless in this condition.

Thank God, there was no human loss at the time of the fire. All the cows and horses got out safely, but sadly, two calves in a pen did not make it out alive.

Did you know that sometimes frightened horses will run back into a burning building because their stable is the only place that they feel secure? Often, owners will blindfold the horses when they bring them out of a burning building so that they cannot see the fire or their stable.

After my tragic fire my owner was told by a neighbor of a farmer who saw smoke pouring out of his barn. Luckily, there was a cistern nearby so the man was able to extinguish the fire with buckets of water before there was too much damage. He discovered that his two sons, age six and seven, were playing "firemen" in the barn. They would scoop up a small stack of straw on the floor, light it and then rush to the fire to put it out with bottles of water. What they didn't anticipate was the possibility that the fire they started just might get out of control. Do you suppose those boys found it uncomfortable to sit down for a period of time? Were you aware that my frame has no nails in it? In the first place it would be difficult to find a sixteen inch nail. Secondly, did you ever try to drive a nail through oak wood?

I was built through a process called timber framing. Each beam is either slotted in a U shape or cut away leaving the center with about four to six inches to fit the slot in another beam. They are called tenon and mortise. Then, a man using a hand or breast auger drills a hole through the sixteen inch beam with the two beams connected together. He then hammers in an oak plug about two inches in diameter through both beams locking them together.

Did it ever occur to you why obesity was not much of a problem in olden days and especially on a farm? The primary reason was that they worked so hard that they burned off the calories. Besides they did not have fast food or junk food to eat.

Did you know that I was known as a Yankee barn because I was built on level ground? The older style barn was known as a bank barn or Pennsylvania Dutch barn It was built with a sloped embankment that went up to the large front doors with the cattle on a lower level such as the Risley-Whipple barn that is mentioned in earlier pages. That design originated in Europe.

That is your history lesson for today.

Cooper Shop Barn

I will bet that you are asking, "what in the world is a cooper shop"? It is a place where wooden barrels or wooden casks were built. Men who built the barrels or casks were called coopers. Cooper is a derivation of the word, "coop" which means pen or cage. Haven't you ever heard of a chicken coop? If you haven't, then you never grew up on a farm.

Wooden barrels use to be a very popular storage item during the time of the early settlers and the development of our country. Neither corrugated cardboard nor plastic had been invented.. Barrels were used for everything, and especially for the transportation of products such as food, supplies, small furniture and liquids, such as whisky, wine and water.

My owner, Jack, had turned me into a cooper shop where he built and sold barrels to ship pottery and dishes that were made in the community of East Palestine, Ohio. Often, there was a lot of steam within my walls since the boards that made up the barrel and the sapling straps used to hold the barrel together had to be steamed to be bent.

His brother, Lester, had lost his job during the Depression so Jack offered Lester a job to gather saplings in the woods to be able to feed and support his family. So Lester set up a tent out in the woods and would gather saplings through the week. He would come back into town each Saturday with as many as a hundred saplings and then return to the woods with food supplies for the following week.

When Lester's son, Ivan, was three years old he took Ivan along with him in the woods. Ivan lived and worked with his father in the woods for three years. In addition to the food supplies that they brought from town, they lived on wild game and wild edible plants.

When Ivan was a teenager he would become so nostalgic for those wonderful years living in the woods and eating wild game that he would sometimes kill, skin and roast a woodchuck over an open fire in his backyard. His mother, being a very wise woman, would not permit him to prepare a roasted woodchuck in her home because woodchucks have a large amount of fat that is strong in odor. Anyone for roast woodchuck?

Throughout his adult life, Ivan loved the woods and all of nature. He believes that it was due to his unusual experience of living in a tent in the woods with his father for three years when he was only a very small child.

The Hooded Rambo

My owner had died and his portly wife took on the task of maintaining the farm the best she could with a few cows and sheep. She had a ram that I liked to call Rambo because of his mean streak of ramming anyone from behind who dared enter his domain, especially, the owner's wife.

Tired of being bruised by his vicious ramming she took a piece of burlap bag, about one foot square with a hem on one side, through which she ran baling string (a farmer's duct tape) and somehow tied that around the ram's neck with the burlap falling over its face.

When the ram bent down the burlap would fall away from its face so it could see the grass to eat or drink from a trough. But whenever the ram would lift up its head to see what was going on in the pasture the burlap fell over its eyes and it could not see. So he never bothered her again.

He was a sight to behold but no longer a threat to anyone's derriere. Talk about necessity being the mother of invention!

Recently, I heard of a tragic situation of a neighbor who had raised this ram from birth and it had never been a threat to anyone. But a ewe in the pasture had been injured so the owner went out to assist the wounded ewe. The ram, apparently wanting to protect the wounded ewe, attacked the owner, ramming him up against some farm machinery viciously again and again.

His wife came out to attempt to rescue her husband and was also attacked by the ram. Both of them were hospitalized. After a month's hospitalization he died from his injuries. She is slowly recovering.

Please be careful around farm animals. They may look so cute and affectionate but you never know what they might do.

Photo By Jan

Halloween On The Farm

I must admit that Halloween is a most exciting time around here. Anything can happen, especially when there are some mischievous boys around. One of their favorite tricks was to run as fast as they could and tackle a corn shock just about where the baling string is tied around the shock. The ride to the ground is as exhilarating as tackling a fullback on a football field but with no bruises, only cushy landings. A more dangerous trick was for them to drag a corn shock to the middle of the road and wait to see what happened when a car driver approached. Usually, they heard a bit of profanity about those darn kids, much to their delight.

I know of one young lad who volunteered each Halloween to provide pumpkins for his church's activities. What the church officials didn't know was that he was stealing them from his neighbor's pumpkin patch. Tell me, was it less of a sin since the stolen pumpkins were to be used by the church?

One of my owners had a son who ran around with five other ornery boys. When six outhouses happened to get knocked over one Halloween night my owner confronted his son as to whether he was involved. The boy admitted his guilt probably with more of a grin of achievement than any reflection of guilt. So, my owner rounded up all six boys the next morning and marched them around the community to reset the outhouses back on their frames. Do you suppose that was as much fun?

Two sons of another owner were with a couple of other boys who leaned a stolen pumpkin on a stick against the front door of one of their neighbors. When the owner of that home opened the door, the pumpkin fell into his vestibule smashing onto the floor. He angrily picked up pieces of the pumpkin and threw them at the laughing, retreating boys.

The most fun for me was when children would turn me into a haunted barn. They had compartments built of bales of hay through which the guests had to crawl with maybe a piece of cow's liver slapping them in the face, or ghostly slimy hands would reach out and grab them. All such action would cause shrieks of surprise and giggles of laughter. After all the children had experienced the haunted barn they usually would have fresh apple cider and donuts. They might even bob for apples in a washtub. Did you ever try it? It is tougher than you think and usually wetter than you might expect.

My one owner used to have a masquerade Halloween party in my barn for the Women's Society of her church. She would never be there as the hostess when the women began to gather. They would only find out which one was her when the masks came off.

Halloween was such a fun time!

Photo By Jan

The Glass Is Broken

I know when you look at me in my deteriorating condition you may feel sadness. It may remind you of your own mortality.

I once had an owner who was conversing with a neighboring farmer. The neighbor said something very provocative. He mentioned an old Chinese saying, "The glass is broken". He went on to explain, "Whenever you pick up a glass, especially a sentimental one, be sure to think to yourself that the glass is broken. Actually, at that moment the glass is totally intact but eventually it will break and be destroyed in some fashion. All material goods and all of life eventually dies. If we are prepared to accept that which we cannot change then we can face whatever is in the future".

I was reminded of mountains and rock that appear to be eternal. Yet, we know that mountains can be turned into volcanic molten lava and rocks can end up in pebbles and sand. Did you know that clay is actually decomposed rock?

The fact that human beings die, along with pets, birds, homes, and even barns is all in the natural order of the universe and creation.

My owner responded to his neighbor by telling him of two terminally ill cancer patients whom he had visited. One was an older woman, probably in her sixties, who had been a devout Christian. She had weekly Bible studies in her home. She had once stood up in his church and announced that she only came to save the members of that congregation. She had put so much trust in God saving her life. But she was obviously dying which put her in the deepest despair of anyone he had ever met. She felt that God had failed her.

That same day he visited a man who was dying in his own home. He was surrounded by loving family members. He was cheerful, as were all the members of the family. He was at peace, trusting in God's care and promises of eternal life. He was ready to accept that which he could not change.

My owner then recalled another terminally ill cancer patient who had cared for a number of the members of her family, including her husband who had died of cancer. Her comment was "others ask 'why me' but I say 'why not me'?"

5 5'07

The Sports Arena

My owner had two young men, Jimmy and Glenn, who grew up on our farm. Through the years I became their primary source of recreation.

In their younger years I became the backboard for their endless playing catch, bouncing tennis or rubber balls against my outer wall. I really didn't mind. Actually, I felt that I was involved in playing catch. They would throw the ball at me and I would always throw it back at them. I loved to throw it back at them beyond their reach and make them chase after it.

When they became young teenagers they installed a backboard and basketball hoop on my main floor center post. With my concrete floor they had their own basketball court that they played on all year long. When they would miss the backboard and it would end up in the hayloft I figured that it was my ball. At least, I made them come after it.

They usually played one on one and sometimes two on two. They might have played more boys but there were not more boys around. I must tell you that I loved their company.

When they grew into young manhood, Glenn moved away. Jimmy had taken up golf. He liked to practice with real golf balls but didn't like the idea of retrieving them out in the field.

So, guess what he did? He started to use my front wall again as a backboard so as not to have to go after the balls.

Would you believe that I caught one of those suckers and even kept it for a few years? It happened to hit a knot hole and stayed there. I suppose you could say that he made a hole in one!

Anyway, it was my trophy to keep for a few years with a number of people coming to admire it and my ability to catch.

What Keeps Me Going

Actually, I have no idea what keeps me going. But my future is not really in my control. (I started to say "hands" and then remembered that I have no hands.)

The elements, especially in Ohio, can bring me down or some human may decide to tear me down to build bigger ones, (as it mentions in the Holy Scriptures), salvaging some of my oak beams and framing, or I could be destroyed by fire, but it matters not to me.

I have had a rich long life filled with joyous memories, activities through the seasons, and struggles that are a part of everyone's life. I had many owners and their families plus the cattle and horses who lived in my stalls. I served them all well.

Now I am so fragile that I cannot hold myself up. Now I am totally at the mercy of the elements or some human's decision. No matter.

Those who loved me do not see me as a pile of rotting boards and pieces of metal roofing. They see me in their memories as I once was and what I meant to them. That is enough!

By the way, isn't that true of aging humans as well?

1 8 '05

Camp Wanake Inpost

I am so a barn! At least, I was a barn until 1945 when the United Brethren East Ohio Conference bought the farm to make it into a Christian campsite. I was built somewhere between 1830 and 1850. I was originally a cantilevered barn which means I had an over-hang. (No, I didn't have a hang-over!) The purpose of the cantilever was to provide greater space for hay storage on the main floor.

The new owners would have liked to have torn me down and built a new assembly building, but they were anxious to hasten the new property into a functional campsite. So, they renovated my cow stalls into a kitchen and dining room and the main floor became an assembly room and sleeping quarters. Eventually, they added a library and stage. Presently, the library is now an air-conditioned series of offices.

They did not like my looking like a barn, so they added a porch with stately pillars and a grand stairwell on my south side. When David Scharr was a junior camper, the counselors threatened to fine the campers a nickel each time they called me a barn instead of an assembly hall. It is interesting to note that David later became the camp director for many years and now his daughter, Julie, is the director.

Some years ago they had acoustical material blown onto the ceiling to reduce the noise in the dining room. It has the color and appearance of globs of oatmeal, so the legend is spread by the counselors to the most naive children that some previous group of campers had an oatmeal fight and they were never able to get it off my ceiling.

I am known as the Camp Wanake Inpost. "Wanake" is an Indian word meaning "a beautiful spiritual place among the trees." It is a very appropriate name for our 250 acres of magnificent hills of trees in southern Ohio. It is also appropriate to say that it is a spiritual place with all its emphasis on the Christian experience and life.

I am very proud of my unusual structure as a barn. I am nearly 90 feet long and 60 feet wide. My joists are rough hewn logs 62 feet long, 20 inches wide, and 12 inches thick that were cut from the surrounding forest. They are oak, American chestnut and black walnut.

Much of the windows, doors, pews, flooring and siding for the first three cabins and myself came from abandoned rural churches that were victims of WWII. So I have been consecrated more than once.

The very first campers paid $10.50 for a week of camping. In 2005 the base cost was $250. Special activities, such as horse back riding are extra.

Generally, about 125 campers, ages 6 to 18, attend each week during the summer. On occasion, they interact, such as in combined dining, vespers or special events when one or two groups work together. Most of the buildings, including parts of me, have been winterized so that programs continue throughout the year.

I must admit that it is more fun and meaningful being part of a Christian campsite than just being an ordinary old barn. I know that I am a very integral part of this beautiful, spiritual place among the trees known in the hearts of the thousands who have been here as beloved Camp Wanake. Thank you, Lord!

Camp Wanake Outpost

I am located about a half-mile south of the Camp Wanake Inpost you know what (I'm not allowed to say "barn") with all its fancy renovations. I still look very much like a barn even though I have more windows than an ordinary barn. I am also involved in the Christian program and am no longer holding cows, horses, or even hay. But, I do own a pet six foot blacksnake who lives in my lower section. Presently, my lower section is only used for storage, such as the many canoes that are used on the adjacent pond.

I partially claim Mazy, a slightly crossed breed, blue eyed, Siberian Muskie. Mazy spent so much time at the camp that the neighboring owner gave Mazy to the camp because she would never stay at home. So now, Mazy spends her days going from one group to another getting constant hugs and snacks, which includes anything edible.

Since my windows are wide open throughout the summer, I also harbor a number of bats who only come out at dusk, so most of the campers don't even know they are present.

I even have sleeping quarters for two counselors who have to experience the hottest room in the entire campsite. Their room is right under my roof. They have to climb a vertical ladder to get to their sleeping quarters.

Would you believe that I have a climbing wall 17 feet high and 12 feet wide on my main floor? It is for the younger children to climb harnessed to a belay safety rope. It is unbelievable to see the number of children who try it and many actually reach the top and ring the bell of victory.

In another part of the campsite there is a three sided, 46-foot climbing wall that has various levels of difficulty. It is interesting to note that more girls reach the top and ring the bell than do boys. I heard one of the instructors explain that the boys burn out before reaching the top because they tend to climb with their arms pulling them up while the girls use their legs to climb. Secondly, he said that girls have much smaller feet so they can stand on much smaller rocks as footholds.

Each summer day my main floor becomes the dining room for as many as 40 kids and counselors with the food brought out from the main kitchen. During the rest of the day I am used for various group meetings, such as worship, crafts, and even a "barn riot" which includes a lot of noisy singing, shouting, and wild modern dancing that shakes the floor to my foundation. Often, they even churn out homemade ice cream the old fashion way using a hand crank, salt, ice and a lot of muscle.

Even though I may not look it, I am actually younger than the Inpost barn because I was built in 1860.

I, similar to the Inpost barn, love how I am being used in the Christian growth of children, youth and adults. Furthermore, I no longer smell like a barn if you get my drift.

18 10'05

Please Be Careful On The Farm

Did you know that the farm has the highest accident rate of any industry within the United States? I do not choose to share with you the fatal accidents or serious injuries that have taken place on the farm because they are too gruesome. I would like to tell you of some of the unbelievable accidents that I witnessed or heard about as farmers shared their stories in my presence.

I remember this young man who was trying to help out my owner by driving the tractor while seeding the field with grain and fertilizer. Unaccustomed to driving a tractor while pulling a grain spreader, the young man made too sharp of a corner that caused one side of the spreader to be lifted off the ground onto the back tire of the tractor. This caused the seed and fertilizer to pour out on the ground. Unable to move the tractor forward or backward, the young man decided to unhitch the short tongue of the spreader from the tractor.

When he pulled that pin the tongue of that spreader flew by his face at lightening speed. Had he been even inches closer, he would most likely have been killed. For a few minutes he stood there trembling with the reality of what could have happened and thanking God for another day of life.

One day John was tossing hay down from my loft. Normally, he would climb down the vertical steps nailed to my center beam. This day the child in him decided to jump down onto the pile of hay that lay on the main floor. Instead of landing on the center of the hay pile, he landed on its sloping side which threw him onto the blades of a manure spreader that was stored there. He actually didn't break any bones but he could have been seriously injured.

A more amusing incident with John was when he had an old tractor sitting on the slope in front of my large doors. Even though he had turned the motor off, he had left it in gear. He was stretched out on his lawn in the cool shade of the big maple tree in his yard taking a snooze. Suddenly, he awoke as he heard that tractor start up. To his amazement that tractor was aiming right for him. Naturally, he arose in a hurry and somehow was able to jump on the tractor and stop it.

Why that tractor took off by itself no one will ever know. Some of his neighbors suggested that it might have been a gremlin who was after John. Do you suppose that John treated that tractor with more respect and tender loving care after that event?

I heard of a farmer who drove his tractor onto the main floor of his barn a little too fast. He didn't know that grains of wheat had been spilled all over the floor. When he hit the brakes, the tractor skidded across the floor as if it were on slippery ice and went right through the back wall, flipping upside down. Luckily, the farmer was not hurt but slightly embarrassed.

Then there was a neighbor of mine whose wife would not permit him to chew tobacco in their home, so he kept his chewing tobacco out in the barn. After losing a certain amount of chewing tobacco over a period of time, he suspected that his hired hand was stealing it. He resolved the problem by lacing the chewing tobacco with horse manure. He never lost any more of his chewing tobacco. One more reason not to chew tobacco. You may never know what it is going to taste like! Didn't I tell you to be careful on the farm? Besides, thou shall not steal!

Grandpa and Grandma Whipple's Barn

I was rather a complicated barn, if I must say so. You see, I lived in Conneaut for many years when there were few barns left in the city. Would you believe that North Conneaut, where I actually lived, was once called "Oklahoma"?

I'm sorry that my memory fails me so I can't really tell you anything about my earlier years as to when I had cows, horses or whatever. That was so long ago. My owners were Andrew and Stella Whipple, who were always known as Grandpa and Grandma as long as I can remember. They were married for over 70 years with Andrew living to the age of 91 and Stella living to the age of over 105.

I only remember becoming a combination barn/garage for the Hupmobile four-door sedan and a chicken coop filled with chickens clucking and a couple of roosters crowing, especially early in the morning. I can't remember when I last had fresh hay in my second-floor loft. Both downstairs and on the second- floor I was a hodgepodge of a little bit of everything from decades past. It was old stuff never used anymore, long forgotten, covered with years of dust.

But the humans, and especially the children, loved to wander through my rooms. They were fascinated with the antique tools, the unique odor of dried leather from an old horse saddle, blue medicine bottles and glass quart milk bottles scattered about, wooden shaft golf clubs in skinny decayed cloth golf bags along with drawers in beat up old furniture that hid a thousand surprises. The odor of musty old hay and straw, mixed with the pungent odor of the chicken coop created an odor that was different from any other barn any one ever visited. You could have brought a blind- folded person into my barn and they would know where they were by the odor alone.

I suppose that I should have been embarrassed by being turned into a chicken coop with no other animals, but I honestly tell you that I enjoyed their happy conversation with each other, especially at dusk when they would roost and softly cluck to one another as they drifted off to sleep.

I no longer exist but I remain in the hearts and minds of those who still remember. They still can remember my unique odor. When they do, they smile.

The Horse Trainer

I am proud of the fact that my latest owner was a race horse trainer and I berthed a number of race horses.

A funny thing happened the first day that she and her husband moved into the farm house. For a moment she thought that she had lost her husband. As he opened my front door he fell backward onto the ground. She raced from the farm house to the barn thinking that he had a heart attack or had broken a leg. When she got by his side he was getting back to his feet.

He explained that he had been startled by a large owl that came swooping out of the barn as he opened my door.

One day while she was taking care of one of her race horses the horse accidently stepped on her foot and literally, tore off her big toe. It was hanging on by a little bit of skin. Her husband rushed her to the emergency room at the hospital and the surgeon was actually able to sew back the toe. Now, it is only a painful memory.

An ironic situation took place on my east wall where you see my louvered window. A spar hawk (sometimes called a kestrel hawk) had built a nest in my louvered window. Notice that almost directly under the window is a doorway. For some unknown reason, someone had left a bucket, half full of water, next to the doorway. Tragically, one of the young fledgling, too young to fly, fell out of the nest, into the bucket of water and drowned. I felt bad for the baby bird as well as its mother.

My owner had raised a heifer from birth. It was kept in a small stall with a four-foot wooden fence. One day my owner entered to find the calf standing outside the fence. They never did figure out how that calf could possibly climb over that fence. I know but I would rather keep them guessing.

Lost Religion

My owner's home happened to be only a stone's throw away so I was able to hear and see many interesting events. One Sunday morning after church my owner's wife and children drove into their driveway. As they approached the house they were shocked and frightened to see ugly blue smoke pouring out of the kitchen of their home.

Racing into the kitchen they found no fire but a horrible stench with thick smoke pouring out of the kitchen stove. Opening the broiler they discovered a not too clean carcass of a big, old woodchuck whose fat was spitting out little, crackling flames and a stench you would not believe.

My owner, who had no idea how to boil water, had gone to his friend's house who had caught a woodchuck in a trap. After they killed the woodchuck, my owner decided that he would try to cook the woodchuck before his wife came home because he knew that she would never cook it. So, he hurriedly skinned the woodchuck and placed it in the oven. Not knowing the difference between the oven and the broiler, he somehow lit the broiler without burning down the house and hoped to cook the woodchuck.

Well, my owner's wife would have been better off if she had never gone to church that Sunday because she practically lost all of her religion in that moment.

The spitting, fiery, smoking carcass was thrown out the back door. Later that day the boys gave the woodchuck a proper burial out back. As you might expect, dinner was late that Sunday-very late!

I don't think my owner ever tried to cook anything again.

Holy Barns

My new owner happened to be a woman minister. She has turned my main floor into a holy place for Bible studies. I carefully listen as they talk about God who I knew, especially when I was growing up in the woods as a series of young trees. I loved Him then and I love Him still.

I look forward to every Bible story that is discussed by the group. They also sing a lot with one young lady playing a guitar. They often pray. Sometimes they will have refreshments along with jovial fellowship. I feel that I am very much a part of this weekly event and enjoy being a holy barn.

Down the road there is a Methodist congregation who were worshiping in the high school while their new church was being built. One of the parishioners owned a vacated chicken coop that was simply being used for storage. So, they cleaned it all up, painted it and even put curtains in the windows. The Women's Society, the mens' group, the youth, trustees, and other small groups met there during the week when the school was not available. You can image the various teasing that the Women's Society heard since they were meeting in the chicken coop.

Have you ever heard of cowboy churches? I heard some farmers talking about them. It seems that they are very popular out west and in the southern states. These are men and women who are not comfortable sitting in traditional churches.

They meet in barns or rodeo arenas or outside with the pulpit and altar sitting on a flat-bed truck. The floor in the barns are usually covered with straw or sawdust so the worshiper can spit their tobacco.

The men and women usually drive up in their dusty, rusty pick-up trucks. They come in their work clothes, which usually means blue jeans, a plaid or denim shirt, cowboy boots and a cowboy hat. They often use the horse trough for immersion baptism.

They do all the things that are done in a traditional church such as singing, but they happened to be cowboy gospel songs. They pray and listen to a preacher who may remind them that Jesus was born in a stable or that Jesus was a real cowboy.

Model T Fords

Most of you people have no idea what farmers experienced with their Model T Fords in spite of the fact that they totally revolutionized the farm as well as all of society. There is some question as to whether or not one could buy an entire Ford from Sears and Roebuck but every part of the Ford was available in their catalog. Fords only came in black with no fancy gadgets or instrument panels on the wooden dashboard. They had two forward speeds, low and high, and one backward speed. The first Ford was produced in 1908 and sold for approximately $900. When the first Ford came off the assembly line in 1914 it sold for $275. Recently, I heard that the 1914 three door Model T Ford was declared by the Board of Directors of the Ford Company as "the car of the millennium." Ray Stanley owns one and gives rides to the residents and their friends at Copeland Oaks, a retirement community in northeastern Ohio.

This one family parked their Ford in my barn but they had a hard time to get it started since they only drove it about once a week to go to town for groceries. So, they would push the car out and hitch a horse to the front of the car. Then they would put the car in gear and walk the horse around in a circle until the car would start. When they got through with their shopping in town they must have started it with a hand crank.

I was located on top of a hill near the road. It was such a comical sight seeing those Fords deal with that hill. Since the fuel pump was not invented, the gasoline was gravity fed from the gas tank. Originally the gas tank was directly under the front seat. To fill the gas tank the owner had to take out the front seat of the car. Later models located the gas tank under the windshield to alleviate the problem of gravity fed to some degree. To get up any hill of any size the driver only had one choice, he had to turn the car around at the bottom of the hill and back it up to the top of the hill, then he had to turn the car around again to continue down the road. It was always so much fun to watch.

The earliest Fords did not have starters but only magnetos. The driver would set the magneto for a certain amount of spark, then the driver or a passenger would go to the front of the car and turn a hand crank. The person cranking the engine was actually turning the four pistons over hoping the magneto would spark to ignite the gas fumes. Occasionally, the pistons would backfire causing the crank to kick, sometimes violently, so some victims ended up with a broken arm or, more often, a sore arm.

There were many horses who were spooked by the Fords coming down the road, especially when they backfired as they passed a horse or horses. When cars were alongside of the road, stuck in a ditch or with one or more flat tires, the most popular response from those passing by was always "get a horse"!

What A Rare Gift

My owner was a veteran of WWI. As an MP at Camp Sherman in Chillicothe, Ohio one of his tasks was to go to each of the barracks to help remove the bodies of any of the soldiers who had died the night before, supposedly from the flu epidemic. It was not until 1938 that the government discovered that there was another epidemic that had swept the US and Europe from 1916 to 1920 known as lethargic encephalitis. Very likely, some of the bodies that my owner helped to carry were afflicted with this mysterious illness that killed most of its victims.

My owner became one of its victims who was lucky enough to survive. But his family never could understand why he slept so much. He tried numerous doctors without any help. Eventually the disease took on all the symptoms of Parkinson Disease. Later, he also lost the control of his eye-lids even though he had 20-20 vision. This may have been caused by his falling out of the back of his pick-up onto his head in a corn field.

He struggled to support his family by owning an antique and used furniture store. In addition, he usually helped his father-in-law with the chores here at my place. He so often would come home emotionally and physically uptight.

As a means of relaxing he and his young son would go fishing in Lake Erie which is close enough to see from my upper window. After supper, the two of them would go and dig a few worms and then go down to Brown's pier where they would cast out their lines. Propping up their rods they would then go onto the beach and build a small fire close to the water's edge.

Then my owner would lie down in the cool sand and his young son would lie down beside him, crawling into his father's arms. There they would talk of many things, while they watched the fire burn down and listen to the waves splash upon the loose rolling stones. Only the young lad would ever watch the poles and would jump up whenever a fish would bite the bait and cause the pole to wiggle.

What a sacred and precious time that was for that young son with his father, a rare, rare gift that few boys or girls ever have the privilege to experience. That young lad grew up to become a Christian minister who had great compassion over the sick, the dying and all who were suffering. How much of that compassionate spirit do you suppose might have come from those sacred moments with his father?

Did You Ever See A Boat Barn?

Well you have now! I have no idea what year I was born but it was possibly in the late 1800's or early in the 19th century. Please observe the height of my large front doors which are about 15 feet high and 15 feet wide. They were designed for taller boats. On the right there is a door that is only about 8 feet tall and 12 feet wide. That was for smaller boats. The small door on the left was for mules or horses who were stabled just inside. Would you believe that I have identical doors on the south side as well?

The most unique part of me is that I happen to have two cross beams on each side of the main floor. The lower crossbeams are the normal beams that hold up the second floor on which boats were stored. About 10 feet above the second floor there are another set of crossbeams even bigger than the lower ones. They appear to be just sitting in mid-air. But, if you look closely there is a smooth round pole the size of a telephone pole that rolls back and forth on top of these upper crossbeams. From the center of the pole hangs a loose chain that would have a block and tackle hitched to it. The round center pole would be rolled back and forth to more easily lift and store boats on my second floor. From various points in my ceiling there were also chains hanging down where blocks and tackles could be attached to more easily store the boats.

Since I was built before motor boats were invented I would have primarily been a boathouse for sail boats and some valuable dories.

In the trash and debris on my first floor there is a remnant of a metal skid that would have been drawn by a horse or mule. Since my location is a half mile from Lake Erie the animals pulled skids or low flatbed wagons from the lake to my location or from my place to the lake. In later years the men drove gasoline driven tractors hauling bigger and bigger boats, including yachts.

After electricity lines were placed on poles along the highway the men bringing the boats to or from the lake had a new problem. They had to have a man riding at the highest point on the boat with a long wooden pole that had a V shape cut at the end of the pole to catch low hanging electrical and telephone lines and lift them above the boat.

One day, one of the workers was riding four feet under a high powered electrical line. He never touched the line with his pole, but the electrical power jumped from the line to the man, knocking him out. Luckily, one of the other workers saw what happened and stopped the tractor. They gave the man mouth to mouth resuscitation until the ambulance arrived to take him to the hospital. Not only did the man survive the ordeal but he still works for the boat company at Browns landing. I don't know if he still rides the boats down the road or not. I wouldn't blame him if he no longer chooses to do that kind of work.

Although it is not too noticeable from my north side, I am rapidly deteriorating on the west and south side. I won't be here much longer unless some kind soul would like to restore me to my former beauty. I must admit that most of my interior beams are mighty solid. Do I have any takers?

By the way, it is so amusing that so many natives of the surrounding community, including some of the grandchildren of my owner, never realized that I was a boat barn. How about that!

3 9'05

Guess!

People who drive by and see me so close to Lake Erie would most likely figure that I must have been some kind of barn used by commercial fishermen to store their nets and possibly some small boats, but they would be very wrong.

Actually, I was a barn to house mules in the 1880's for the Kelly Island Limestone and Transport Co. in Marblehead, Ohio. The mules pulled small racks on skids loaded with limestone that was mined from the quarry to the crushers or to the kiln where they manufactured lime. These products were shipped by train throughout the United States and Canada.

In the early 1900's they installed narrow gauge railroads and purchased Shay steam locomotives from the Lima Locomotive Works that were called "dinkeys" because they were so small. Isn't that a terrible name? Even though they were small, they were mighty in their ability to pull heavy loads of small gondolas loaded with limestone. From 1898 to 1923 they purchased thirty five Shay engines, two hundred wooden gondolas and four wheeled dump cars.

I can't remember when they no longer kept the mules with me. For a long time I was left vacant. But then some enterprising officer of the company decided that I could be made into four apartments. That was the most exciting and delightful time of my life with all those families and especially all those kids. But since 1985 I have been vacant again.

Now, there is a faded "for sale" out front. Rumor had it that someone plans to buy me and tear me down in order to put up some expensive condominiums. They call it progress but I call it tragedy.

Heina Lato

Heina Lato {pronounced hay-la-tow) means that "I am a hay barn in Finland". I was never built for cows and horses. My primary purpose was to store hay or any kind of foliage. I am sitting in the middle of an oat field in Kitinoya, Finland. Ed and Naomi Pollari from Ohio, USA visited me this summer and took my picture. Ed has cousins all over Finland and comes to visit them every once in a while.

I am not really alone out here in the middle of this field. There are other Heina Latos on this farm as well as many on the adjacent farms. We kind of enjoy being by ourselves. It is a very tranquil life style. You, in your busy lives, ought to try it sometime.

Traditionally, the primary purpose for placing hay barns on various fields was safety from fire. If one of us happened to catch on fire our owner would not lose his whole hay crop. Secondly, it was very convenient to have storage nearby rather than all in one barn.

By the way, the pile of wood leaning up against my wall is simply a storage place for firewood that will be used to heat my owner's home.

I can't remember how old I might be but it could easily be over 150 years old. Don't you think I look pretty sharp for being that old? I know that you are questioning my age because you do not see many 150 year old barns. The reason that wood does not rot in Finland is because we have very low humidity both summer and winter.

I still have a grain binder in one end of my shed that is over 100 years old. At one time I was used as a grain dryer. My owner would bring in sheaves of wheat or oats and lay them upon my rafters. Near my entrance there was a big concrete box filled with rocks. My owner would build a fire under the rocks so the heat would rise to the rafters and dry the grain.

When the grain was dry my owner would bring the sheaves to the floor and beat the stuffings out of the sheaves with a flail to separate the grain from the straw. A flail was a long wooden handle about 8- foot long with a leather swivel attached to one end. On the other end of the leather swivel hung a 3- foot wooden bar. Afterward the straw would be lifted aside into a storage bin while the grain would be scooped up, taken outside in an open container and winnowed by throwing the grain into the wind so the chafe would blow away and only the grain would be left in the container. This was the same way that grain was threshed in Biblical times.

10 9'05

The Renaissance Barn

How do you like that for a classy name? Not only do I have a distinguishing name, I think that I look pretty sharp for as old as I am. I was built in 1880 which makes me 125 years old. I got the name because I happen to be the repair and storage shop for both the Renaissance Retirement Community and the Links Golf Course. The Links Golf Course, which is a very popular public golf course, is available free to all the residents of the retirement community. All the revenue of the golf course goes to the Eliza Jennings Senior Care Network of which the Renaissance retirement community is apart. I was spruced up when the Renaissance Retirement community was established in 1990. Don't you think that I look pretty good for my age?

The history of the farm, of which I was a part, goes back much further than 1880. My original owner, John Hall, was born in 1838 in England. I have no idea what may have happened to John's father, but his mother, with John and his six sisters, crossed the Atlantic, sailed up the Hudson river to Albany and traveled over land to Buffalo. There they boarded another ship and sailed to Cleveland, Ohio arriving in North Olmsted in 1847.

The various members of the Hall family owned large amounts of farmland in both North Olmsted and in Dover, which is now known as Westlake. They were such a prominent family in the community that the present day Columbia Road was known as Hall Road until 1980. The road upon where the Renaissance Retirement Community and the Links Golf Course are located is named John Road in memory of John Hall.

The Hall family built a very beautiful brick mansion that sat just north of me. For many years it was the club house for the Links Golf Course. When the Renaissance Retirement Community was built, the mansion had seen its better days and it was torn down. A new club house was built across the road. Where the mansion used to sit there is now housing for assisted living residents. A beautiful gazebo is also there for the residents to enjoy.

Life brings many changes but life goes on. Thank God we have our memories and the richness of our history.

A Streetcar Barn Named Desire?

I truly was a car barn in Marblehead, Ohio for the interurban electric cars that ran all across the state of Ohio from 1888 to 1931. Prior to 1888 they had horse drawn or mule drawn cars in the large cities that were slower than molasses.

By World War I there were 2798 miles of track in the state of Ohio. This was over a thousand miles greater than any other state. Every town in Ohio that had 10,000 or more residents were connected together with interurban street cars. Actually, the streetcars and railroads brought on the demise of the amazing Ohio canal system that used horse drawn or mule drawn boats.

The interurban cars held 20 to 58 passengers. The smaller cars were primarily used in urban areas. The earliest cars were made of wood, but the later cars were made of metal. The smaller cars operated on a single 50 HP motor, while the larger cars would have as many as four 50 HP motors. Slapping at their railroad competition they often published ads that read:

"NO SMOKE - NO CINDERS- NO DUST".

They had all kinds of cars, such as open cars for summer, private cars for the very wealthy, baggage cars, utility cars, refrigerated cars, cranes, snowplows, and various other pieces of mobile equipment.

Most of the cars traveled as single coaches but sometimes they would run in tandem or have a baggage car or trailer hooked onto the back. In 1920, when the World Series of Baseball was played in Cleveland they combined as many as nine cars to handle the crowds.

Some of the cars, like many that came to Marblehead, were double ended. The operator could operate the car from either end so that it was not necessary to turn the car around.

The cost of a ride was usually 15 to 60 cents. Luggage, including bicycles, tricycles, and baby carriages cost 25 cents each.

In our area the interurbans were mostly used to bring tourists to the Lakeside Campground or for people wishing to take a boat to one of the islands in Lake Erie.

Just as the Ohio canal system bit the dirt due to streetcars and railroads, so the interurban street cars died from the development of the automobile, busses and trucks. The great depression was the straw that broke the camel's back. Most of the interurbans were gone by 1931.

I hung around until a few years ago when I was torn down to provide a parking space for a ferry boat company. I suppose you might call it progress. Oh well, I had my day. It was a great ride as long as it lasted!

Photo By Jan

Dead Or Alive

You may see me in this deplorable condition and consider that I am more dead than alive. But what defines life? Granted there are portions of my being that have changed in their process of decay, but if a board in my structure has been eaten away by wood bores, or termites and the remnants of that board have become fertilizer for new generations of plant life and eventually, human need, is that part of me dead or merely transformed into a new life?

Doesn't the scripture promise us that we will be transformed and have a new and different body? Are you so presumptuous as to claim that our Lord was speaking only of human life and not of the future for animals, birds, fish, insects, trees and even barns?

Is it possible that you can observe the transformation of the caterpillar into the gorgeous butterfly and not recognize that such transformations take place in all forms of life? At what point do you consider that the caterpillar actually died or is it still alive even in its transformation in the life of the butterfly?

I heard some humans talking that the purpose of life is to glorify God no matter in what condition you may be or how old you may be. Did it ever occur to you that I might live by the very same purpose?

The huge oak logs that make up my basic structure sought to glorify God, their Creator, by being the most beautiful and strongest oak logs that they can be. They are still holding up my frame. Eventually, they may be torn down but not destroyed. They are very apt to become beautiful exposed joists in a ceiling or a fireplace mantle. Are they dead or alive?

The pine boards that make up my siding never tried to be an oak tree but the best and fastest growing pine trees in their young lives that they could be. Even the nails that hold the siding to my timber frame never tried to be wire but remained strong and firm fasteners of my boards to my frame as nails. Does the fact that they eventually turn to iron rust make them dead or alive?

Every portion of my various parts sought to be perfect as possible just as our Creator in heaven planned for us to be, fulfilling our Creator's will. Isn't that your primary purpose too?

10 9'05

Horses Should Not Make Pigs Of Themselves

Did you know that some horses will gorge their grain that is put in a feed box at the front of their stable? When that happens they will get sick (excuse the expression) as a dog and develop colic. To avoid that problem my owner would put a few round stones the size of your fist in the grain box. This forced the horses to eat more slowly as they had to move the stones around with their nose or tongue to get all the grains in the box. When the horses were through eating you can rest assured that there were the same number of stones in the box as there were before they began eating.

But cows are different. They don't seem to know when they have foreign substances in their mouths. When they are eating long strands of hay they are apt to consume any pieces of metal that happens to be hidden in the hay.

One of the cows in my stable happened to swallow a small section of barbed wire. How the veterinarian figured that out I will never know. Apparently, it happens enough times that the vet carries a magnet in his medicine bag. He pulled out the magnet and check through the stomach area from the outside of this cow and found the location of the barbed wire. He then proceeded to give the cow a series of local aesthetic shots and then cut a hole into the side of the cow as well as the stomach lining. He then proceeded to reach in with his hand and removed the barbed wire. Then he sutured up the cow's stomach and side. I guess that it wasn't as bad as trying to pass that barbed wire through the natural channels. Ouch!

There was another time that a vet had to be called because of a breached, unborn calf who had died in the womb of the cow. Once again the vet had to cut a hole in the side of the cow to extract the dead calf. The cow survived the ordeal to live a long life but it never again could bear another calf. I didn't blame her.

5 8 '05

I Speak Of History

Would you believe that I am located about a thousand feet from one of the main intersections of Westlake, Ohio? Westlake has become one of the most prestigious communities on the west side of Cleveland. This is where the most wealthy and most prominent people of Cleveland come to live. For instance, it has become the home of many of the famous players for the Cleveland Browns and the Cleveland Indians, as well as many other modern day celebrants. So, I stand proudly among their fabulous homes, knowing that there is no other structure as old or as outstanding as I am.

Actually, I have no way of knowing really how old I might be, but I suspect it was around the Civil War. I was built by George Cooley, who became the father of Ohio Farm Bureau for which he received a medal of honor from President Coolidge. He received a second medal of honor from President Coolidge for his service to agriculture in our country later in his life.

Our community was called "Dover" until 1940. The fact that there was another community in Ohio that had the name "Dover" caused a great amount of confusion for the post office. So, in 1940 the community voted to change their name to "Westlake".

They named the road that is in front of me Dover Center Road to remember our heritage. I think that it is a very appropriate name since I am also a part of the heritage of Westlake.

From my spacious ground floor, the present owners, who are still part of the Cooley clan, are retired and sell produce they grow, just as Uncle Charley did.

There have been several additions to my original structure but if you were to walk in my original section you would see my true age. I think they should make me into a museum for Westlake in respect for the pioneer, Uncle Charley. What do you think?

I Was Built With Marriage Marks

I know that sounds weird but when I was built my owner used marriage marks in the assembly process. Marriage markers have nothing to do with marriage ceremonies or the institution of marriage but they were a system of assembling a barn. The framing would be laid out on the ground. The carpenters would drill out two rows of three holes adjacent to each other on the vertical beams where they wanted to insert cross beams. Then they would chisel out the hole that they had created with hand chisels and a hammer to make a very neat rectangular hole. This hole is called a mortise. Then they would saw and chisel the end of a cross beam to fit snugly into that mortise hole. This projection at the end of the cross beam is called a tenon. They would have to make a similar tenon on the other end of the beam to fit into the next vertical beam. In order to avoid mixing up the ends once the builders would start erecting the beams the carpenters would mark the mortise and the tenon with the same Roman numeral. This was known as a marriage mark.

My only guess as to why they called them marriage marks were that the two parts of the framing became one and generally remained together for the life of the barn. If only the marriage marks of humans could be as permanent.

After the cross beams were erected and secured into place with the vertical beams the carpenters added one more way of securing the frame of the barn. They would drill a hole through where the two beams were joined together and drive a wooden peg through the hole. The peg was the length of the width of the vertical beams which were usually 16 to 20 inches square.

Did you know that all barn frames were hand hewn until about 1910? That means that all the barns built before 1910 were built using only hand tools and primitive equipment. I am proud to say that I was built before 1910. Don't you wonder how those pioneer farmers were able to do so many amazing things?

Photo By Jan

The Barn Consultant

My owner invited Chuck Whitney of Mt. Vernon, Ohio to come examine me. I was very proud to have such a distinguished guest. My owner has been talking a great deal about his coming. Chuck actually has a newsletter called "The Barn Consultant". He has traveled more than 300,000 miles in 10 different states and visited more than 600 barns.

The fact that Chuck is well beyond the age of retirement means that he is not only an expert in his knowledge of barns but he grew up with many of these old barns that he explores. He claims "I have experienced all the changes in farming from the 1920's to the present day."

He is not only interested in the condition of every barn he inspected but he is also interested in its history. He does a thorough inspection of each barn he visits, checking the original siding, the timber framing on the inside and the specific uniqueness of each barn. He often consults with owners who are planning to make some changes in their particular barn for some specific purpose that they have in mind.

Rarely does he visit a barn unless he is invited by the owner or is given permission by the owner to look over his or her barn. He never visits a barn without his reliable jack-knife that he often uses to cut a splinter from some timber to check the kind of wood that was used in that particular structure. It is not unusual to discover barns with several different types of cross-beams because the original builder would use whatever was available in the wood lot nearby.

I am pleased to announce that Chuck considered me in very good shape for my age. Most important was the fact that when he cut a splinter from my frame he discovered that many of my beams were made from American chestnut, a species that has been extinct for more than 50 years. That made me just feel a little taller that day.

He had photos of numerous barns that he has inspected. One unusual photo is a barn that was three stories high. The first floor had the stables, the second floor was for grain plus machinery, and the third story was used for hay and straw lofts. It would not have been practical except for the fact that it had been built in the side of a hill.

Another photo showed a barn that was completely built with stone walls. The walls were made with slabs of limestone. The walls all tapered from 3- foot wide at the bottom and 1- foot at the top. I can only guess that the original owner needed to get rid of some limestone.

He also had a photo of a barn under construction that was assembled with two jin poles.

Before the days of cranes to lift beams in place the builders would plant one or more telephone like poles firmly into the ground with cables attached to the top of the pole. The jin pole had to be taller than the highest part of the barn. They would attach one end of the cable to a portion of the framing and the other end would be pulled by horses or strong men to lift it in place and secure it.

I heard Chuck tell of one daring man who stood up on top of a jin pole. His wife told him to stay there until she went into the house to get a camera. She took the picture and he somehow got back down safely. No one should try this at home.

GARDEN
12 5 '05

I Am A Civil War Survivor

Excuse me if I look a bit run down, but you have to remember that I am over two hundred years old. My present owners, Henry and Judy Bonebrake, claim that their Georgian style brick home, which is only a stone's throw from me, was built in 1840. As the important carriage barn to the home, I have to be close to the same age.

The fact that I am located in Hagerstown, Maryland and only 14 miles from the important and tragic Civil War battle known as Antietam, you can realize that I sit on a very historic site. I was fortunate to survive the Civil War. When the house was renovated a number of years ago, parts of a Confederate uniform and a Civil War sword were found. Even today, family members occasionally pick up artifacts from the war in the surrounding fields, such as bullet casings.

It should not be forgotten that the battle at Antietam was the bloodiest single day's battle in the Civil War. General McClellan lost 12,410 men of some 69,000 engaged and General Lee lost 13,424 men of 52,000 engaged for a total of 25,834 men in one day. This seems totally unbelievable when the primary weapon of war was the single shot muzzled loaded rifle. Of course, other weapons were also in use. We should moan over the losses of both the North and the South.

I am embarrassed by that ugly chicken coop which was attached to my frame years after I was built, but of course, I didn't have a say in the matter. Just to my north there is a 10 by 10 foot square brick smoke house, which was the primary way of preserving meat in the early days and would still be functional today if people still smoked meat.

How do you like the look of my pink face with my big dark eyes and cute, round nose? No mouth is shown because barns are not expected to eat or even talk. I may not eat anything but I love to talk to anyone who will listen to me and I have seen a lot through the years with my big dark eyes.

A few years ago we had a lot of excitement around here.. Two convicts escaped from the MCI penitentiary in Hagerstown. The next morning, Judy's son, Kevin, and his friend, decided to go horse back riding. When they went to the barn that was just east of me (before it was torn down), they found the horse blankets that were put under the saddle were missing from their usual place. Looking around they found the blankets in another part of the barn, along with towels and paper articles that were from the penitentiary.

Immediately running back to the house, the boys informed Henry who called the State Police. In a matter of minutes there were streams of State Police carrying shot guns combing all the buildings, sheds and homes as well as helicopters flying low overhead hoping to find the fugitives. Even I was searched! The convicts actually escaped, but I suspect they eventually were caught. It was sure exciting around here for awhile.

Tales Of Conneaut Creek

There is so much I could tell you pertaining to my location on top of the bank of Conneaut Creek, right next to the Nickel Plate railroad tracks. I remember the time that my owner's foolish eight- year- old boy decided to crawl out on the single track railroad bridge that stands a hundred feet above the creek. There were men working on the bridge who had safety islands that hung alongside the bridge for them to stand on when a train passed by. They were about 200 feet out on the bridge when they spied this young lad, who was too scared to walk upright on the wooden ties, crawling on his hands and knees out to them. What he didn't realize was that there was a train coming out of the railroad depot less than a mile away. One of the men raced across the bridge, grabbing the boy as he ran, and got to the safety of the bank just before the train arrived.

Another time this boy and his sister were standing on the bank near the tracks when they heard a man crying for help who was sitting on a large rock in the middle of the creek. They ran to their parents who phoned police and an ambulance. Later, the newspaper reported that the man had been pushed from the roof of a box car by a railroad detective just as they were passing over the bridge. Miraculously, the man landed in the water and only broke his leg.

When this same boy had reached the 3rd grade he, along with eight other 3rd and 4th grade boys, would jump the fence at noontime at school and go skinny dipping in Conneaut Creek. Their problem was that they did not always get back to school in time for the opening of the afternoon classes. So, the teacher warned them that they were not to do it again or else they would face serious consequences.

Accepting such a challenge, they all went one more time and did not get back until recess in the afternoon. When they returned, the women teachers sent all of their children outside and came to view the act of corporal punishment that was applied by a male principal who worked at a school five miles away. He used a piece of rubber hose which was applied to each boy's bottom as the boy stretched over the desk. Needless to say, that was the last swim for the boys.

When that young boy grew into manhood, he still did not do the smartest things. Having returned from military service, he and his family had a picnic at Farmington Dam on Conneaut Creek. Without checking the depth of the water he dove off a log into the water which was much too shallow. He hit the shale bottom with his head, coming up with blood streaming out of his face. He had a large cut on his forehead and had broken his nose. On the way home, they stopped at a doctor's home who bandaged up the wound on his face and plugged his bleeding nose.

When he got home, he was lying in bed when he suddenly realized that he could no longer open his jaws or talk. All he could imagine was the possibility of lock jaw. His mother anxiously called in their neighbor who was a chiropractor. The moment the doctor saw the young man he began to laugh. He announced that the young man had dislocated his jaw. Climbing on to the bed, he placed each of his knees on the boy's shoulders, stuck his fingers into the young man's mouth and pulled with all his might while his father held down his head. Suddenly, the jaw snapped back into place. The young man was so excited that they couldn't get him to stop talking.

Conneaut Creek has so many wonderful memories for my owner's family and for me as well.

The Unbelievable Four's

I am not going to hide the identity of that boy any longer. It is time the truth is out. I was located at the back of their little house on Nehemiah Street (now called Washington Ave) in Ashtabula, Ohio. The Whipple family lived here for a few years until the 1929 Depression. While they lived here the youngest boy, Paul Jr. reached the age of four. You have heard of the "terrible two's" but have you ever known of the "unbelievable four's?" Paul Jr. was one of them.

One day, Paul Jr. and his buddy, Bobby (also age four) saw the fire engines down at the end of the street racing to a house that had a roof fire about a block and a half away. So they ran to see the fire. They were not satisfied to watch with the rest of the crowd. The front door was open so they entered, unnoticed by the gathering crowd.

They bounced up and down on the living room soft chairs, then wandered through the dining room into the kitchen. After inspecting the kitchen they started back through the dining room when some of the now water-soaked ceiling plaster started falling. Not yet seeing any fire, they struggled up the stairway over the large four inch canvass/rubber hoses to the top of the stairs. There they were surprised to see the blue sky through the hole in the roof and the flames that the firemen were trying to extinguish.

When one of the firemen looked down and saw those two boys they, for the first time, heard of number of words they had never heard before as the firemen grabbed both boys, dragged them down the stairs and pushed them out the front door. Apparently, the crowd never saw them because they were looking up at the fire in the roof. Paul Jr. saw his mother standing in the group of spectators so he sauntered over and took her hand. She asked him, "where have you been?" He never answered her until he was in high school. (Please do not consider his mother as irresponsible. Life was much simpler and less dangerous then, except for four year olds.)

I wish I had time to tell you more of his many antics but I will leave that for another time. I will tell you of an unwittingly serious accident he did cause that came back to haunt him fifteen years later.

The Fowler family, who lived across the street, were all getting into the family car. Paul Jr., in an attempt to be helpful, slammed the back door shut right on the youngest daughter's thumb. She was screaming in pain, and blood was squirting all over the place. The family piled back out of the car. The mother wrapped the hand in bandages until they could get Donna to the doctor. Frightened and feeling very guilty, Paul Jr. raced for home. It was not long after that episode that the Whipples moved to Conneaut and the Fowlers moved to Arizona.

After serving in WWII, Paul Jr. was home when he was notified that Donna was in Ashtabula. So Paul Jr. made arrangements for a double date with another couple. At 18 years of age, Donna had developed into a beautiful young lady. Sitting in the back seat of the car together, Paul Jr. noticed that the thumb on her right hand was severely deformed. He asked her what happened to her thumb. Her shocking answer was "you ought to know, you did it."

I suppose he probably deserved it but he always moaned, "I could have spent the rest of my life without ever knowing that I had deformed a beautiful girl!"

(I hate to think that some beautiful tree sacrificed itself to make this paper and I left it empty. This has nothing to do with barns but they understand and approve my using this space)

My Dream Journey

My dream journey would be to travel back to my youth and childhood. There I would relive the joy of the innocence of youth, such as exploring the mirco-world of tiny ants and grubs under a piece of decaying bark or splash barefoot through a shallow creek lifting rocks to try to capture a darting crayfish.

There I would climb into the limbs of an apple tree in full bloom, surrounded by the smell of the sweet nectar of the blossoms, watching the honey bees lazily fly from one blossom to another undisturbed by my presence. I would lie safe in the limbs of that tree, dreaming fantasies of adventures, amazing feats of strength and skill, and heroic deeds that I would accomplish someday when I grew up.

There I would dig some worms and go fishing with my father at twilight on the shores of Lake Erie. We would cast out our lines from the pier that jutted out into the lake. Then we would return to the beach near the water's edge to build a small fire. There we would lie down on the cool sand watching the wood burn and listening to the rhythmic lapping of the waves upon the beach. There I would lie in my father's arms, talking about many things, often ignoring the fish lines that were only an excuse for being there.

There I would walk with my mother through our woods as she would bring to my awareness the song of a bird, or point out a Jack-In-The Pulpit emerging from the brown leaves of the winter past. As we walked upon the spongy, ground ivy smelling its pungent cedar like odor I would know that we were walking on Holy Ground.

There I would laugh and play with my brother and sisters with imaginative trains made from tipping kitchen chairs on their side. I would slide down the steep hill in the wintertime on a piece of cardboard. There I would play marbles on the living room rug inside a ring of string and we would sing for long periods of time with Mom at the piano playing old songs and beloved hymn tunes.

There I would experience the first breaths of love in a gracious smile from the girl next door, or the glee of dancing eyes, or the excitement of holding hands, or the unforgettable shy first kiss.

There I would go to church with my family to greet all our friends and extended church family. I would have the fun of Daily Vacation Bible School where I would paste, cut and color various artwork which I would proudly bring home to show my mother. There I would go to Youth Camp at Lakeside to know the spiritual mountain top experience that was so moving I would not want to come home at the end of the week.

But since such a journey is impossible I will thank God that I have the blessed joy of just remembering!

Paul Whipple